# Moving Heaven and Earth

# Moving Heaven and Earth

## PhysWizz Gravity

**Alexander Greener**

**Ros Greener**

Copyright © 2019 by Alexander Greener

All rights reserved. This book or any portion thereof may not be reproduced or used in any manner whatsoever without the express written permission of the copyright owner except for the use of brief quotations in a book review.

www.PhysWizz.co.uk

ISBN: 9781075717864

*"The important thing is to not stop questioning."*

*Albert Einstein*

# Chapter One

Today was the last day of school term and Alex was on the bus home. He was daydreaming of all the adventures he would have in the holidays.

The next week would be especially exciting because Uncle Tom was coming to stay. Uncle Tom was his Mum's uncle and lots of fun. He was always coming up with ideas for crazy adventures and inventions.

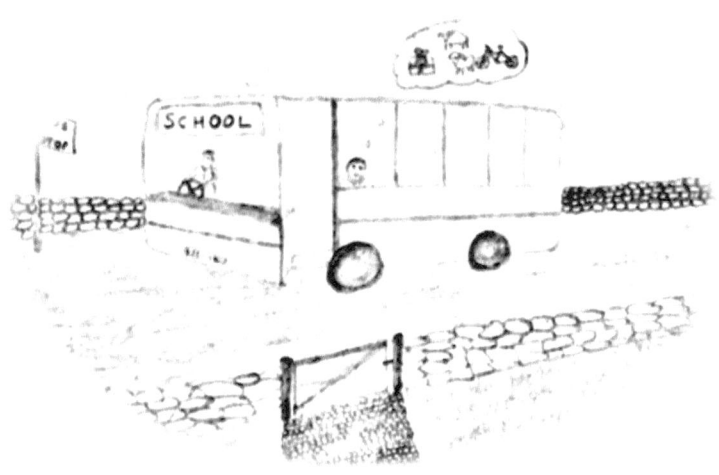

Alex wanted to be an inventor when he grew up. He often spent whole weekends constructing new machines. Sometimes they worked, sometimes they didn't, but he was already thinking about the next thing he wanted to make.

Alex jumped off the school bus as soon as it stopped near their house and ran all the way home.

"Is Uncle Tom here?" asked Alex as soon as he came through the door.

"Yes, he's just down the garden," said Mum.

Alex raced into the garden and jumped into his Uncle's arms. Uncle Tom gave him a big hug, then spun him round and round until they were both laughing and dizzy and Uncle Tom fell over.

"I have another surprise for you," said Mum following him into the garden. "Sue has been called away on a last-minute business trip, and she asked if Sam and Joe could stay with us for a few days."

Sue was Mum's friend from university. She was often travelling with work and sometimes her children Sam and Joe came to stay whilst she was away.

"Oh yes! That's fantastic," said Alex, jumping up. "When are they coming?"

"Sue is just bringing them over now. So, in about ten minutes. Why don't you get changed out of your school clothes before they arrive?" said Mum.

Alex rushed upstairs and was just coming back down in his T-shirt and jeans when the doorbell rang.

"Hello Alex!" shouted Sam and Joe rushing up to him and giving him a big hug. "We've come to stay for a few days, it's so exciting."

Sam was nine and her younger brother Joe was five. They both went to Clover Hill school in Dewbridge, the same school as Alex. Sam hoped to be an explorer when she grew up and always carried a little bag with useful items for exploring wherever she went. Joe liked to follow Sam on her explorations, usually asking constant questions as they went.

"What shall we do?" asked Sam.

"Do you want to play space rockets before dinner?" suggested Alex.

"Yes! Yes!" shouted Sam and Joe.

Alex had a stomp rocket from his last birthday. The sort that fires foam cylinder rockets with a foot pump to stamp on. Alex liked to pretend they were blasting off into space and counted down from 5 to 1 before stamping on the foot pump and launching the rocket skyward. They had occasionally ended up in a nearby tree or bush, which then meant throwing something a bit more solid up to knock them back down.

Alex, Joe and Sam each tried to get their rocket the highest. Sam managed to get hers as high as the biggest tree in the garden, although a gust of wind took it into the neighbouring field causing a stir of interest amongst the sheep there. Luckily, they managed to retrieve it before it was nibbled.

"How hard do you think we would need to stamp to get the rocket into space?" asked Joe. "It would have to be really hard."

"I think we would need more force than the air pump can give," said Alex. "And foam is maybe not the best material for a real rocket," added Sam. "The ones I have seen look like they are made of metal and they blast fire out from the rocket, rather than being pushed up from a ground-based launcher."

"Do you want to try with a water rocket?" asked Uncle Tom, coming to join them. "It is not quite the real thing, but it does have its fuel on board like a real rocket."

"What's a water rocket?" asked Joe.

Uncle Tom opened his mouth to explain, but Sam interrupted. "I know! A water rocket is just a plastic bottle, you put some water in the bottle, then seal it with a rubber bung pushed in the neck of the bottle.

You pump air into the bottle through a needle hole in the bung. When the pressure inside is strong enough it pushes off the rubber bung and the water flies out, which shoots the rocket into the sky. A bit like a real rocket, but with water instead of fire."

"Do you want to try?" said Uncle Tom, putting a little plastic rocket on the ground. "This is one I made earlier. It already has the water inside. It just needs the air adding."

Alex attached the pump and they all counted down as he pumped in the air. They had just got as far as 5 – 4 – 3, when the rocket suddenly whooshed upwards, soaking Alex and splashing Joe and Sam.

The rocket went much higher than the foam rocket. Joe ran to collect it and bring it back. "Me next!" he said.

Uncle Tom poured in some more water and fixed the bung and pump in place. "Off you go."

5 – 4 – 3 – 2 the rocket was already in the air. Sam and Joe squealed as the water soaked them.

They took it in turns until they were all drenched and the smell of cooking wafted towards them on the breeze.

"I think you lot better get changed before dinner," said Uncle Tom.

After they had finished eating, Uncle Tom asked them if they would like to play human rockets.

"What do you mean?" asked Alex.

"I mean, why don't you three make a human rocket?" said Uncle Tom.

"Sounds dangerous!" said Sam.

"Sam you can be the first stage rocket, Alex you are the second stage and Joe, you can be the top of the rocket."

"Here are some tables I adapted earlier with your Mum's help," said Uncle Tom, wheeling over three tables, with large wheels screwed onto the bottom of the legs. Alex's mum pushed the rest of the furniture in the kitchen to the side.

"Sam and Alex, if you each lie on a table, with your arms stretched out and hold the table in front of you. Joe, you lie on the front table and hold a little rocket – the Launch Escape System (LES). This is just in case Sam blows up or there is another disaster on launch. The LES would jet away from the rest of the rocket, taking the astronauts, who are in the top part of the rocket, clear from the rest of the rocket and the launch site.

Now when we take-off Sam will push away as hard as she can on Alex's table, to propel Alex and Joe off towards the lounge, as she goes in the opposite direction towards the patio doors.

Alex will then push away on Joe's table, so Joe shoots into the lounge and Alex goes back into the kitchen. The lounge represents Space and the kitchen is the atmosphere and Earth. Are you ready?"

"Yes," they all shouted back.

"5 – 4 – 3 – 2 – 1 blast off!" they chorused.

Sam pushed away Alex and Joe who started moving towards the lounge. Then Alex pushed away, accelerating Joe towards the sofa, where Uncle Tom was waiting to catch him.

"That was great fun," said Sam, "can we do it again?"

"Yes, of course," said Alex's mum. "Why don't you swap places and see if it's better to have most weight and strength at the bottom or top of the rocket."

They spent the rest of the evening shooting up and down the kitchen on the wheeled tables, trying to see just how far and fast they could propel their human rocket. They all agreed Joe made the best top part of the rocket.

"OK, my little rockets, I think it's time for bed," said Alex's mum, noticing the clock. "I've made up three beds in Alex's room, which should be nice and cosy for you all."

# Chapter Two

When Uncle Tom went up to say goodnight, they asked him if he could tell them a story.

"Would you like me to tell you about a man who actually travelled into space, all the way to the Moon, and back again, in a rocket?"

"Yes, please," they all said together.

"His name was Neil and he worked for NASA, which is the American space exploration agency. Neil trained for many years before being selected to become an astronaut on what was called the Apollo 11 mission to land on the Moon.

To get to the Moon you need travel very fast, so you are not pulled back down by the Earth's gravity. This means the astronauts needed to reach a speed of over 25,000 miles per hour. The motorway speed limit is 70 miles per hour in the UK, so a rocket travels over 300 times faster.

Neil and two other astronauts, Buzz and Michael were strapped on their backs flat on the floor of the rocket ready for take-off. "

"Why were they on their backs on the floor?" asked Sam.

"If they were in any other position, the force of acceleration when the rocket took off would crumple them and the blood would get dragged downwards into the lower part of their body so they might pass out.

If you have ever been in a car or on a rollercoaster when it suddenly accelerates, you may have felt pushed back against the back of your seat. Being in a space rocket taking off is like that but the pressure is much stronger.

The Saturn V Rocket was used to launch them into space and then toward the Moon. Neil had a radio to speak to the team at ground control whilst the rocket's systems were being checked to make sure they were functioning as expected. Then he waited nervously for take-off.

He could feel his heart pounding in his chest with a combination of fear and excitement, as well as pride.

After what seemed like hours, the noise of the engines started and count down began. 5 – 4 – 3 – 2- 1 and Lift Off!

The noise and vibration from the engines were overwhelming as the rocket moved faster and faster upwards. Neil felt like two grown men were sitting on him, squashing his chest and body.

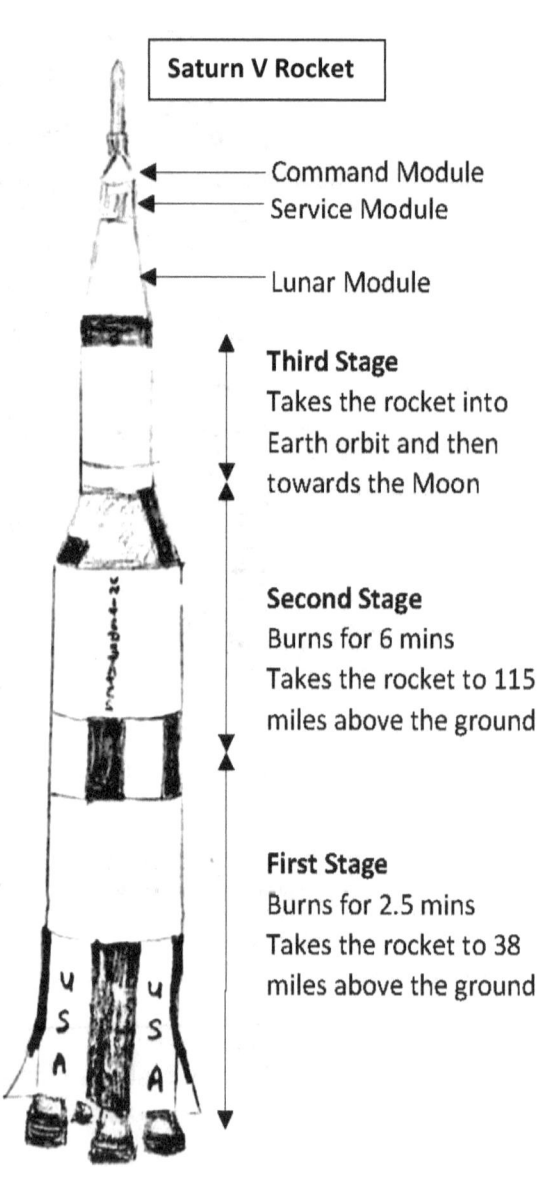

**Saturn V Rocket**

- Command Module
- Service Module
- Lunar Module

**Third Stage**
Takes the rocket into Earth orbit and then towards the Moon

**Second Stage**
Burns for 6 mins
Takes the rocket to 115 miles above the ground

**First Stage**
Burns for 2.5 mins
Takes the rocket to 38 miles above the ground

There was a loud bang as the first stage finished burning its fuel and was thrown away from the rocket.

The pressure on Neil's chest slightly eased until the second stage fired up and the rocket was accelerated high above the earth. When the second stage had burned all its fuel and fell away from the rocket, Neil felt the pressure release on his body. He undid the straps around him and floated free in the tiny Command Module with Buzz and Michael. He felt amazing, floating above the Earth. All three astronauts laughed with relief now the take-off was over, and they were on their journey towards the Moon.

Neil found the weightlessness great fun. Although they had only a small space to move, he could spin around in the air and "fly" from one end of the space craft to the other. "

"Did he lose weight just by going into space?" asked Joe. "My mum is always talking about losing weight."

"Well yes, he lost weight, but he didn't lose mass," said Uncle Tom. "The two are quite different, although on Earth we often use them to mean the same thing. Weight is a measure of the pull of gravity on us and it changes depending on where you are, but your mass stays the same."

"Ah OK," said Joe, "so when we weigh ourselves, we are actually measuring the strength of gravity?"

"Yes, that's right," said Uncle Tom. "Gravity depends on the mass of your body and the mass of the Earth and how far you are from the Earth."

"I get it," said Alex, "so since the Earth doesn't change mass and we are on the surface of the Earth we can use weight as a measure of our mass as well."

"So how do you measure mass without gravity?" asked Sam

"Imagine you hang something by a string, said Uncle Tom, "you can measure mass by how fast it moves if you give it a push. If you lift it up and let it rest in your hand, that is the weight you are measuring."

"So in space, you would still need to push something just as hard to make it move, but that would be the same in all directions, up, down, backward or forward," said Sam

"Exactly," said Uncle Tom.

"What happens next in the story?" asked Joe.

"Neil looked out of one of the windows on the spacecraft and saw the Earth behind them. Like a colourful jewel in the blackness of space. He was overwhelmed with its beauty. He thought of his family and hoped he would get home again safely.

After checking the controls again, Neil went to get something to drink. Without gravity, he couldn't pour a drink into a cup or pour the contents of the cup into his mouth. Instead, the water was in a prepacked pouch with a straw. But it felt wonderful to have a drink and relax after the stress of the launch.

During the training, Neil had been told it was important not to allow bits of food and drink to get into the air, as they would float about and it would be hard to filter them out so all food and drink was carefully packaged to avoid crumbs or droplets escaping.

Going to the toilet was a different experience as well. Neil had a bag with a suction hose for a wee and a bag that stuck to his bottom for a poo. They definitely didn't want that floating around in the tiny spacecraft!

As they made their way to the Moon, Neil thought over and over about what he would need to do when he got there and what he would say. No one had ever walked on the Moon before, he was going to be the first person, and he was very excited. But before that, he would need to successfully land on the surface of the Moon.

It took a few days to reach the Moon. He was almost used to the weightlessness now. Sleeping was still difficult. It felt very strange being strapped down to sleep. He could hear the whirr of the air circulation system. He woke several times from dreams of something going wrong with the Moon landing. Sometimes they crashed, or couldn't reach the surface of the Moon, or lost power. It was going to be high risk, and on the Moon, without any oxygen to breathe and with temperatures below freezing, he wouldn't survive long if it didn't go to plan. They were a very long way from home."

"How far away is the Moon?" asked Alex.

"On average about 380,000 km or 240,000 miles. The distance changes slightly as it goes around the Earth," answered Uncle Tom.

"Wow! That is a long way," whispered Joe

"As they came close to the Moon, they adjusted the rocket into an orbit around the Moon."

"What is an orbit?" asked Joe.

"It means they moved fast enough and close enough that the gravity of the Moon caused them to travel in a path circling around the Moon. The Earth has lots of satellites in orbit around us, including the ones that beam down information for our satellite TV and satnavs, as well as the international space station and the Moon itself.

Neil and Buzz prepared to take the lander (named the Eagle) down to the surface, whilst Michael kept the main Command Module in orbit ready for their return.

They shook hands and hugged and hoped to see each other soon.

Neil and Buzz climbed into the Eagle lander and separated from the main spacecraft. As they descended to the Moon, Neil fired the engines to slow them down, they looked for a safe place to land without too many boulders. The original site planned was too uneven and the lander stirred up too much dust to see through and safely land. They only had a small amount of fuel to find somewhere else. Neil directed them to a flatter area and brought the lander safely down onto the surface. It was a success. The Eagle had landed.

They adjusted their space suits, checked their oxygen packs and opened the door of the lander.

As he stepped onto the surface of the Moon, Neil said "That's one small step for man, one giant leap for mankind."

And it was an amazing achievement for mankind. We had made a space rocket capable of leaving the Earth, travelling to another object in space and landing on it with humans living on board.

It was strange walking on the Moon. The gravity is 6 times less than on Earth, so Neil had to wear special heavy boots to keep him held down."

"Why is the gravity so weak on the Moon?" asked Joe.

"Is it because the Moon has a lower mass?" suggested Sam.

"Yes, the Earth has 80 times more mass than the Moon," said Uncle Tom.

"How many Moons would fit in the Earth?" asked Alex.

"The diameter of the Earth is about 4 times that of the Moon, but the volume of the Earth is 45 times greater, so you could fit 45 Moons into the Earth, with a bit of squashing and squeezing," said Uncle Tom.

"But why is the gravity only 6 times weaker, if the Moon has 80 times less mass?" asked Sam.

"Because the Moon is smaller. The gravity on the surface decreases with distance from the centre. Although the mass is much less, you are closer to the centre of the Moon," said Uncle Tom.

"Neil and Buzz collected a few samples of the Moon to take back to Earth. They put an American flag where they landed, then they climbed back into the lander. They stayed on the Moon to eat and sleep. Then fired the engines to leave the surface and return to join Michael in the main Command Module spacecraft, before heading home.

Neil was only walking on the Moon for just over 2 hours, but they were the most amazing hours of his life."

"Did he make it home again?" asked Joe.

"Yes, they all travelled back to Earth safely and landed in the Pacific Ocean as planned where a US navy ship was waiting to pick them up," said Uncle Tom.

"I'd love to go into space," said Sam. "It sounds amazing!"

"I think I'm quite glad to be tucked into bed on Earth," said Joe.

Alex was already asleep and dreaming of flying to the Moon.

# Chapter Three

The next morning was warm and sunny with a cloudless blue sky.

"Can we have a picnic today please Mum?" Alex asked at breakfast.

"I'm sure I can put something together. Would sandwiches and carrot sticks be OK?" replied Mum.

"Sounds good, where's Uncle Tom?" asked Alex

"I'm not sure, he went out earlier, but he left this note for you all," said Mum.

Alex took the note and read it out.

> Dear Alex, Sam and Joe
>
> This is your first challenge....
>
> Lose weight and gain weight without changing your mass!
>
> Yours, Uncle Tom

Alex, Sam and Joe looked at each other puzzled. "How can you change weight without changing mass?" wondered Sam.

"We go into space?" suggested Joe. "Like Neil in the story last night."

"Not sure we can manage that today," said Sam, "but I'm sure we'll find out. Is there anything else on the note?"

"Oh yes," said Alex, "there's a second page with a map." He put it out on the table for them all to look at.

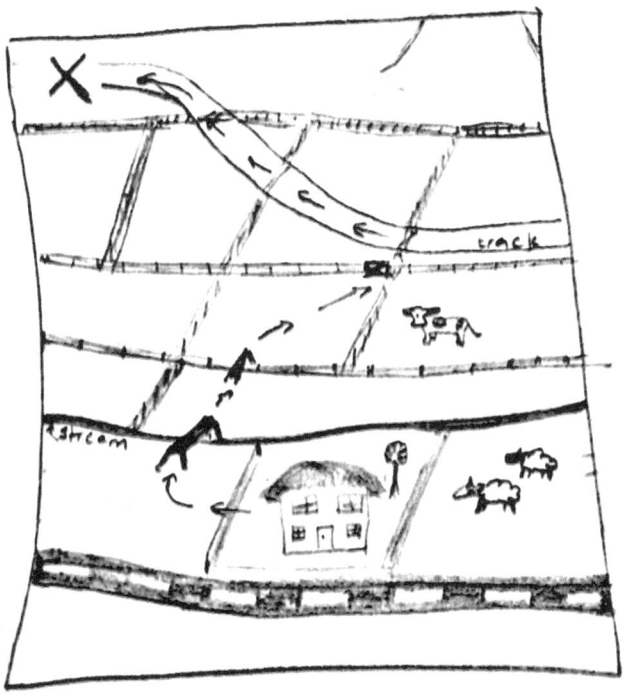

"Well there's the house," said Sam, "and that's the field out the back, so I guess we follow the trail marked on the map. Sounds exciting, let's go."

Mum handed them a rucksack with some food and drink, and they set out.

"This is fun," said Joe, "I wonder what we are going to find. What do you think would make us change our weight?"

"Well I guess if weight is the pull we feel from the Earth and if don't change our mass, then either the Earth changes mass or we move closer or further away from the Earth," said Alex.

"But Uncle Tom mentioned when a rocket is launching to go into space the astronauts feel much heavier. So maybe if we accelerate upwards or downwards, we would feel heavier or lighter," said Sam.

"So, if we jump up and down we are changing our weight?" asked Joe.

"Hmm," thought Sam. "The astronauts' weight can't really have doubled on take-off, if anything they should have been lighter as they got further away from Earth's surface. They just felt heavier because the rocket was accelerating upwards. A bit like being on a rollercoaster when it drops suddenly, our weight doesn't really change, but we feel for a second like we are floating. "

"Or the feeling we have sitting in a car when it suddenly accelerates," added Alex.

"Yes, exactly. So, it changes how we feel, but doesn't change the pull of gravity on us," said Sam. "Astronauts feel weightless in orbit around the Earth, but they are in orbit, so there must still be a pull of gravity on them. But for some reason they don't feel it. But also, they don't fall back to Earth, the gravity keeps them in orbit, like the Moon."

They were now at the far side of the field and climbed over the wall.

"Where next?" asked Joe.

"We need to go diagonally across this field and join the track, then follow that for two more fields. Th.en we should be almost on the X mark," said Alex

"So, we need to change the mass of the Earth, which could be tricky, or we need to get closer or further away?" said Sam returning to the challenge.

They walked on in silence, lost in their thoughts.

"Why don't they fall back to Earth?" Joe broke the silence.

"Who?" asked Sam.

"The astronauts in orbit," said Joe.

"They don't take lots of fuel into space, so they can't be actively moving away from the downward pull the Earth," said Alex.

"Then they must be falling, but they don't hit the ground," said Sam.

"Why?" Asked Joe.

"I'm not sure," said Alex. "Maybe they are just moving so fast that when they fall, they fall around the curvature of the Earth. So they circle around the Earth."

"Like a conker on a string," said Sam. "When you spin something heavy on a string, you can feel the pull of the string, but the conker goes around in a circle. The conker is always trying to fly off, but the string pulls it into a circle. A bit like the gravity pulling the astronauts into orbit. If you let go of the string the conker would fly off in a straight line."

"Yes, and if it wasn't spinning fast the conker would get pulled in by the string, just like gravity pulls things back to Earth," added Alex.

"But why do they feel weightless?" asked Joe.

"I guess, because they are in constant free fall, and everything near them is also in free fall. There is no air in space to make them feel they are travelling at a high speed, so instead they feel like they are floating," said Sam. "We only really feel gravity because something is pressing against us, like the ground under our feet or the chair we are sat on."

They were now almost at the X mark. "I know where we are going," said Alex. "It is the old mine."

The area was once full of old coal mines. They had all closed long ago, but there was one kept open for people to visit. It was run by Mr Mitchell, the husband of Alex's class teacher.

"Hello Alex!" said Mr Mitchell as he saw the children come around the corner. "Have you come for a trip down the mine?"

"I guess so," said Alex, looking at the map again.

"Excellent, well we are just about to set off now on the tour if you would like to join us."

There were a few tourists gathered outside. Already in their hard hats.

Mr Mitchell gave Sam, Joe and Alex their hats to put on.

"Right everyone, let the tour begin. Follow me and mind your heads."

The children followed the others down the tunnel. It was pleasantly cool, although a bit dark for Joe's liking. Sam passed him a torch to light the way.

The adults had to stoop as the roof was quite low.

"More designed for a pit pony than people," said Mr Mitchell.

They carried on following the tunnel down, deep into the Earth.

"We are now some distance below sea level," said Mr Mitchell. "This is where the coal seam started. Some miners would have to walk up to an hour to get to their coal seam each day, depending on which mine they were in. They worked in almost complete darkness, feeling for the coal and didn't leave the mine until their shift was finished. Which meant they ate, drank and sometimes went to the toilet where they worked."

"Urgh!" said Joe.

"Do you think we have changed weight down here?" whispered Alex. Not wanting to disturb Mr Mitchell's talk. "It must have something to do with the challenge."

"Well we are closer to the centre of the Earth so maybe the gravity will be stronger?" Suggested Sam.

"But then there is less Earth underneath us, so the gravity would be weaker," countered Alex.

"So maybe we are the same weight?" said Joe.

"But if we went right to the middle of the Earth," began Sam.

"We would fry in the process?" joked Alex.

"I mean, if we could get to the centre of the Earth, there would be no gravity because the Earth would pull on us the same in all directions. So, if we get closer to the centre, we should be getting lighter as we are moving towards a position where we would be weightless," said Sam.

"But gravity would have pulled the heaviest parts of the Earth to its centre when it formed, so most the mass of the Earth will be in the centre. And the land must be lighter since it floats on the surface. So if we are just a little under the surface the gravity should increase as we are closer to most of the mass," said Alex.

"So we are a little heavier then?" said Sam.

"I think so," said Alex, "we have completed our first challenge!"

Mr Mitchell was still speaking about the history of the mine and life of the miner.

"It was a hard life, with a high risk of accidents and few safety precautions. But compared with other jobs at the time it paid well, and they weren't short of boys and men willing to work. Even young Alex here could have got a job down the mine. Boys from 5 years upwards would have been spending hours every day in the pitch black, hewing the coal to power the nation's industry.

Thankfully things have moved on, but I hope you all enjoyed the tour. Now let's head out to the light again."

They all trundled out back up the tunnel. The warmth and light increasing as they made their way back to the mine entrance.

And outside waiting from them was Uncle Tom.

"So, you completed challenge 1 then?" said Uncle Tom. "Here is your first reward," and he handed something to Alex.

Alex looked at it carefully. It was a puzzle piece with lines and shapes on.

"What is it?" asked Joe. "It looks like a piece of a puzzle."

"I think it is," said Alex. "There must be more we need to collect."

"Yes, there are," said Uncle Tom, "but first let's go and find somewhere for a picnic."

## Chapter Four

Uncle Tom led the way along the path towards Pen Hill.

"Before I forget," said Uncle Tom, "I must hand you your second challenge."

Sam took the envelope and opened it.

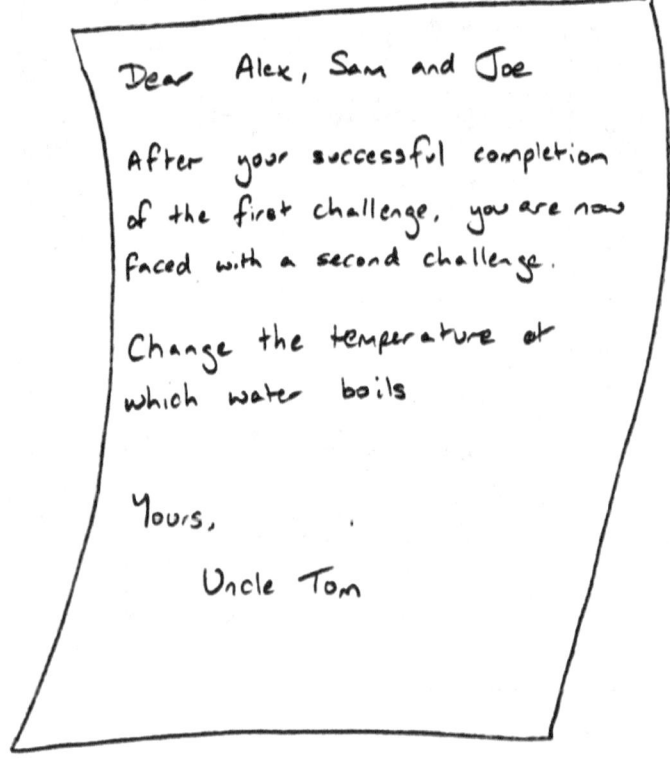

Dear Alex, Sam and Joe

After your successful completion of the first challenge, you are now faced with a second challenge.

Change the temperature at which water boils

Yours,

Uncle Tom

"But first let's find a nice place for lunch. They head up the path to the top of Pen Hill.

Walking up hills always reminds me of a friend of mine, Nick, who climbed Mount Everest. Although Mount Everest is much taller than this hill, or any other mountain I have ever climbed."

"Wow!" said Sam, "what was it like."

"It was really tough," said Uncle Tom, "several of the others who started out with him headed back down before they reached the top. When you climb up a mountain, you are climbing up through the Earth's atmosphere and towards space. As you get higher, the air gets thinner."

"Why does the air get thinner?" asked Alex

"When you stand on the surface of the Earth all the air above you - between the top of your head and space - is weighing down on you, pulled down by gravity. That is the air pressure. But as you climb upwards, there is less air above you to press downwards, so the air pressure gets less and less. If the pressure is less, it means the air molecules spread out more. We often say the air gets thinner. So every breath you take has less oxygen. For each breath we would take at ground level, you would need to take three breaths at the top of Mount Everest to get the same amount of oxygen.

Nick spent several weeks on Mount Everest preparing by climbing up and down the lower sections of the mountain and waiting for the right weather conditions. There are only a few weeks each year when the winds are lighter, and it is possible to climb to the summit.

He had climbed many mountains before, but Everest is the highest in the world and he knew many climbers die each year trying to reach the top.

Nick set off after breakfast along with several other climbers and local Sherpa guides. They carried their food, clothes and oxygen for the journey. They reached Camp 1 on the afternoon. It was hard work but they had all climbed this section several times whilst training in the last few weeks. However, this time it felt different. They were on their way to the top. The next few days would be tough, but Nick was very excited.

They left Camp 1 shortly after breakfast. After only a few hours he noticed his legs feeling more tired than usual. He knew it was just the atmosphere getting thinner and that he was breathing less oxygen. He was very glad when they stopped for a rest.

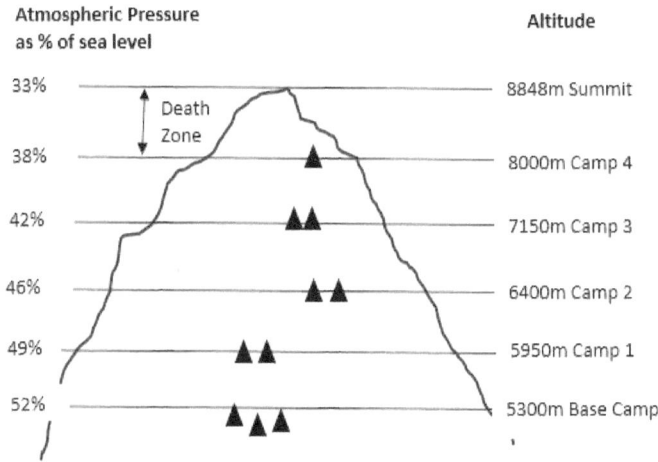

They pushed on again to reach Camp 2. Nick sat with the other climbers eating their dinner and looking back at how far they had walked. It was a long way down, but it was still a long way up to the top. Nick already felt a slight headache and was very tired. He worried if he would make it.

He wrapped himself up in his sleeping bag and went to sleep. It was a restless night. Nick kept waking up and he dreamt of being too tired to reach the top and having to turn back. He had spent so long preparing for this climb, he was determined to succeed.

He was up early. Many of the other climbers had also not slept much. They were keen to get going. After a quick breakfast, they set off again.

Nick's legs were getting more tired. It was much more of a struggle to climb and he could feel himself breathing harder and faster than usual. Nick didn't feel hungry or thirsty, but he knew he had to eat and drink to keep going. It was also noticeably colder the higher he went, and it was hard to keep his hands warm and functioning."

"But hot air rises, shouldn't it be warmer further up?" asked Alex," and you are closer to the Sun."

"That is a good question," said Uncle Tom. "Yes, warm air does rise, but gases cool as they expand. Since, air expands as the pressure drops the higher you go and the more the air expands, the colder it is.

And yes, you are a little bit closer to the Sun, but it is only a very small amount compared with the distance from Earth to the Sun. But the atmosphere does protect us from much of the harmful radiation from the Sun, such as X-rays and Ultraviolet as well as absorbing some of the heat. Because the sunlight has passed through less atmosphere high up a mountain, it is more intense, and it is very easy to get sunburnt.

Nick was glad to stop for the night. He was feeling very tired and a little sick but forced himself to have something to eat and drink. Then went to sleep. He had another restless night.

The next day was a struggle as they climbed up to Camp 4. Breathing was much more difficult, and Nick used some of his oxygen to keep going. This was their final camp before the climb to the Summit. Camp 4 is on the edge of the death zone. At 8,000m above sea level, humans cannot survive for a long time without oxygen.

Nick ate and tried to rest ready for the final climb. He was feeling exhausted and very worried. He had an uneasy rest, then they set off just after midnight in the pitch blackness of the night. 4 out of the original 10 climbers in his group had decided to return to base camp. They were feeling too unwell and did not want to risk the final climb.

Nick had an oxygen pack for this section, but it was still extremely hard work to move and breathe. He knew he had to keep focus to ensure he didn't fall, and he had to keep moving to get to the summit and back in the day.

He did make it to the top and back. But to be honest it has not inspired me to climb Mount Everest. However, it does make me value our atmosphere. It lets us breathe, keeps us warm and blocks harmful radiation. It also destroys meteorites if they come too close."

"Yes, shooting stars," said Alex.

"That's right, they burn up in our atmosphere and we see the burning streaks of light, "said Uncle Tom.

"And it is gravity that means we have our atmosphere," added Sam.

"Absolutely," said Uncle Tom.

"Well now we're at the top of our little mountain that is Pen Hill, and thankfully with lots of air still above us, shall we have a little something to eat?"

# Chapter Five

Uncle Tom opened his rucksack and spread a picnic blanket out for them to sit on.

Alex handed out the little parcels of lunch his mum had prepared for them and cartons of apple juice.

They sat admiring the view and thinking of how different it would be if they were at the top of Everest.

"I wouldn't fancy trying to eat my lunch whilst barely being able to breathe," said Sam. "Or feeling so tired just moving my arms to get the food to my mouth."

"Or being freezing cold," added Joe, feeling the warmth of the Sun on his skin.

"Maybe we should think about the next challenge," said Alex. He took the paper out of his pocket and read it out loud again. "Change the temperature at which water boils."

"I am not sure where to start with this one. I thought the boiling point of water is always 100°C," said Alex. "Maybe we need to think about what makes water boil."

"You need to heat it," said Joe.

"Yes, it needs heating until it boils, so why would you need to heat it less or more sometimes?" asked Alex

"Maybe if you are high up the heat behaves differently, like the hot air rising," said Joe.

"But it is the water that would need to behave differently, rather than just the surrounding air. What happens when water boils?" pondered Alex.

"The water molecules get so hot, they leave the liquid and fly off into the air as gas molecules," said Sam.

"Maybe if you blow on the water," said Joe. "If it is windy, clothes dry much quicker and that is the water molecules turning into gas."

"But if you blow on water, it cools the liquid, whereas if water is boiling all the molecules are turning to gas," said Sam.

"So what else would mean the water molecules could either fly away from the liquid more easily or with more difficulty?" asked Alex. "Maybe if the air above the water is pushing down more it would need more energy for the water molecules to escape and if there was less pressure the water molecules would more easily escape so less heat would be needed."

"There's less pressure high up a mountain, so maybe if you climb a tall mountain, the water would boil at a lower temperature," said Sam.

"Is that it, Uncle Tom?" said Alex.

"Yes, you are quite right, that was excellent teamwork. Yes, at the top of Everest the boiling point is about 70°C. Still quite warm, but not really hot enough for a good cup of tea," said Uncle Tom. "Here's your puzzle piece prize." He handed Alex another puzzle piece.

Alex look at it carefully, it looked a little like the one they got earlier after the mine. He put it safely in his pocket.

"Shall we head back home and see what your mum has been up to today?" said Uncle Tom.

"I'm sure she'll still be sat at her computer working," said Alex.

"Well maybe we should cook the dinner then, shall we have a barbeque?" suggested Uncle Tom.

They walked back in the late afternoon warmth and stopped at the village shop to get some food for the barbeque.

"Good day?" asked Alex's mum when they got home.

"Amazing!" said Alex, "we've tunnelled into the centre of the Earth and climbed up Mount Everest into the sky."

"Wow, that certainly does sound amazing," said Alex's mum, giving Uncle Tom a questioning look.

"And we're cooking the dinner," added Joe

"Even better," Alex's mum said.

# Chapter Six

"You all look very tired," said Uncle Tom as Alex, Joe and Sam settled into bed.

"Would you like me to tell you another story?"

"Yes, please," they said.

"This story is about a boy who was very accident prone. He was always dropping things, tripping over and scraping his knees."

"That sounds like Harry in my class at school," said Joe. "He's always got a scrape or a bruise."

"This boy's name was Barney," said Uncle Tom. "And he decided he would do something about it.

One night when he went to bed, he wished really hard that gravity would go away.

He knew without gravity he would never drop things or fall over again.

He closed his eyes and wished as hard as he could.

When he woke in the morning, he threw back the bedclothes and jumped out of bed as usual.

Except it wasn't as usual, the bedclothes floated in the air and when he jumped, he banged his head on the ceiling, then stayed floating in mid-air."

"The gravity had gone!" exclaimed Joe

"Yes, his wish had come true. He was floating in the air like an astronaut in space. He was so happy! He could spin and twist in the air like a gymnast and had fun flying across the room from one wall to the opposite side and back again. It was the best feeling ever.

He could even lift up his wardrobe and all the furniture in his room and just leave it pushed up against the ceiling."

"He would have still needed a bit of force to make the furniture move, wouldn't he?" asked Alex.

"Yes, he would, but just a little to get it moving. They only needed to move slowly.

There was a glass of water next to his bed, he lifted it to his mouth to drink, but no water came out. Lifting the glass away the water stayed floating in the air as a sphere. Barney put his mouth to the water and sucked some liquid into his mouth. Some of the drops broke away and he had fun trying to catch them all.

He opened his door and made his way downstairs to get some breakfast. It was quite tricky moving, he had to pull himself along by grabbing the stair rail and wall. Without the gravity pressing his feet to the floor, he had no grip under his feet to walk. He floated over to the cupboard and went to pour some cereal into his bowl. He tipped up the carton, but nothing happened.

Looking out the window towards the street, he saw people, dogs, cars slowly floating away from the ground. Without gravity, they had no ability to get back down.

Barney had read enough about rockets to know if he wanted to propel himself in one direction, he would need to propel something else in the opposite direction. Sadly, he didn't have a rocket pack, but instead, he filled his rucksack with big books he could throw to change direction. He opened the kitchen door, jumped up to the top of the house, then pushed his feet off the roof towards the centre of town.

He glided over the town and across to the river. He could see that the water was no longer flowing beneath him. He floated along the path of the river towards the sea. It was a bizarre sight. Boats were floating out of the water. There were round drops of water floating into the air above the sea. There were no longer any tides or waves. The sea was rapidly evaporating as the air pressure dropped."

"Why was the sea evaporating?" asked Sam.

"Without the air pressure pushing downwards, the gas water molecules could easily fly away. If the air pressure was very low, the sea would boil."

"Does that happen in space?" asked Alex.

"Yes, if you threw a cup of water out of a spacecraft it would boil, then immediately freeze since it is very cold in space.

Barney was finding it hard to breathe. He knew the atmosphere was disappearing into space. It was getting cold and getting darker as the Earth moved away from the Sun.

Barney started to cry. His tears pooled in his eyes. What have I done? He said and tried to throw himself on the ground in despair.

He woke up with a start as he fell out of bed. He opened his eyes and laughed out loud. It had all been a dream. He was very glad that gravity is here to stay and after that, he didn't mind when he fell over."

Alex, Sam and Joe slept soundly dreaming of floating around in a world without gravity.

# Chapter Seven

Alex woke first and gently pulled back his duvet, just checking it came back down to the bed and didn't float away. He didn't like the idea of a world without gravity, although it could be fun to float around and spin in the air.

He saw Sam open her eyes. "What shall we do today Sam?" he asked. "Do you think Uncle Tom will have more challenges for us? We have two puzzle pieces, there must be more and then we could complete the puzzle."

"Let's go downstairs and find out," said Sam.

Alex's mum had put everything out for breakfast.

"Did Uncle Tom leave anything for us today?" he asked

"Yes, there's an envelope over here with your names on."

Sam ran over and tore open the envelope.

She read the note inside.

> Dear Alex, Sam and Joe
>
> After your successful completion of the second challenge, your third challenge is to make something heavier than water float.
>
> Yours,
>
> Uncle Tom

They rushed through their toast and cereal keen to get started.

"We could take a few things down to the stream and see what floats," said Alex.

"Good idea," said Sam, "let's go and find things we can take."

"This glass is heavy, maybe we can take that?" suggested Alex.

"Could you pick things that aren't breakable please," said Alex's mum. "What about metal, that is heavier than water. "

"OK, what is there in the kitchen we can take?" said Sam.

"We could take a pan?" suggested Alex.

"What else is there?" asked Joe.

"We can take a metal spoon and fork and this metal bowl too," said Alex.

They put all the things into a bag and headed down to the stream below the house.

The stream was one of Alex's favourite thinking places, especially on a hot day. He could sit in the shade of the trees with his feet in the water, listening to the gentle babbling noises of the water and watching for any little fish swimming by. It was shallow enough to paddle easily across, although the stones could be slippery, and he had more than once fallen into the water and soaked his clothes.

The three of them found a nice area of grass near the edge of the water and laid out all the items they had brought.

"Which ones should we try first?" said Joe.

"I think the spoon and fork will definitely sink, so let's try those first," said Sam.

"OK." Joe picked them up and dropped them into the water. They made a satisfying splosh and dropped straight to the bottom.

"They definitely sink," said Joe.

"Shall we try the bowl next," asked Alex. He placed the bowl on its base on the stream surface.

It floated easily and drifted along with the current downstream. Sam started getting up ready to follow it, but the bowl had already tipped over and was sinking as the water rushed over the brim.

Sam took off her shoes and socks and paddled into the stream to retrieve the bowl from under the water.

"Let's try the pan," said Sam, "placing it gently on the surface of the water. It travelled a short distance before it tipped over. The lid fell off and the pan sank as it filled with water."

She paddled in and emptied out the pan. "Maybe we should try with the lid stuck on." She dried the edge of the pan and lid on her sleeve and took out some sticky tape from her bag. She wrapped the tape several times around the lid, sealing it to the pan.

"Let's try that again." She placed the pan into the water. It tipped slightly but floated off downstream.

They stood up and walked along the stream bank as the pan floated along next to them.

"The pan sinks when it fills with water, but floats when it is empty," said Joe.

"It is not completely empty," said Sam, "it is full of air."

"But why does that make it float. Is it like a hot air balloon, lifting the pan upwards?" asked Joe.

"I suppose it is similar, air rises through water," said Alex.

"Because air is much lighter than water," added Sam

"So if we fill the pan with air, even though the pan itself is heavier than water, the air is so light that the combination of the pan and air inside is lighter than the water. So, it floats. We made metal float," said Joe.

"Just like a boat," said Alex. "They are giant floating bits of metal filled with air."

"Is that like a plane?" asked Joe. "They are large bits of metal that float in the sky."

"But a plane would drop out the sky if the engines turned off," said Sam, "so it doesn't really float. A boat will always float, unless it takes on water, like our bowl."

"Like the Titanic did," said Alex. "It hit an iceberg which put a hole in the base of the ship so the water could get in and the whole ship sank."

"The pan is floating away," said Sam, noticing the flash of sunlight on metal as it disappeared around a bend in the stream. They broke into a run to catch up. "I don't think Mum would be pleased if I came back without her pan," said Alex.

When they were level with the pan, Sam waded in, grabbed the pan and lifted it out the water, putting it in the bag with the bowl, knife and spoon.

"That was just in time," said Alex. "The stream joins the river just around the next bend. We would have had to swim if it got that far."

They wandered on towards the river and turned to follow it downstream.

Then Joe noticed Uncle Tom sitting on a bench by the riverbank. "Hello," he shouted.

Uncle Tom waved. "You guys make a great team. Well done, you have completed Challenge 3."

"Do we get another piece of the puzzle," asked Joe.

"Of course," said Uncle Tom. "Here it is." Alex took the puzzle piece and put it in his pocket with the others.

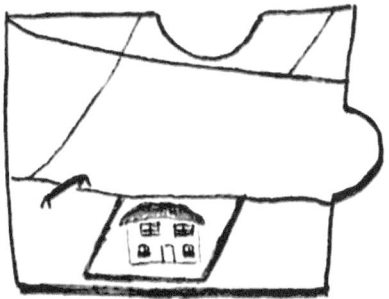

"Shall we keep following the river and see where it goes?" said Uncle Tom. "Let me carry that bag, it looks a bit heavy."

# Chapter Eight

They walked along the river, watching the water flowing along beside them.

Joe thought he saw a fish near the surface, but it had gone when the others turned to look.

They could hear the seagulls, and as they rounded the next bend, the sea was visible in the distance.

"Are we going to the seaside?" asked Alex.

"All rivers flow to the sea eventually," answered Uncle Tom.

"Why do they all go to the sea?" asked Joe.

"The water flows downhill, so the sea must be at the lowest point," said Alex.

"The land must be higher than the sea," added Sam, "otherwise the sea would wash over the land and it would be submerged."

"But some rivers flow into lakes," said Joe.

"Yes, that's true. But then the lake fills up until the water can run out of the lake downhill again and the river continues onwards," said Sam.

Sam loved the sea. "Let's run we're almost there."

They reached the seaside village of Seaburn. It was a small village with a lovely sandy beach. Today it was bustling with families and children excitedly carrying buckets and spades heading towards the beach.

"We may not have buckets and spades," said Uncle Tom, "but we do have a pan, bowl, fork and spoon. Let's see what we can build with those."

"But first, is anyone else a bit hungry. Would you like some chips?"

"Yes!" they all shouted.

They sat on the beach, eating their chips watching the sea.

"Why is the sea salty?" asked Alex.

"That's a good question," said Uncle Tom.

"The rivers aren't salty, and the water in the sea must come from the rivers," said Joe.

"And the rain," added Sam, "but that's not salty either."

"But the rivers wash away the rocks as they flow over them, which is why we have river valleys, and those bits of rocks must flow to the sea," said Alex. "If some rocks are salty, the water will dissolve the salt and it will flow to the sea."

"When the water in the sea evaporates it leaves the salt behind. We did an experiment in class with saltwater. When it is left in a jar for the water to evaporate the salt crystals are left behind," said Sam. "So, the sea water evaporates and the salt from the rivers gets concentrated in the sea."

"As well as lots of other minerals and rocks washed into the sea by the rivers," said Alex. "The salt dissolves but some of the other rocks will just sink to the bottom of the sea as sediment."

"That's right," said Uncle Tom. "The rivers carry lots of minerals including salt into the sea."

Now they had finished their chips it was time for sandcastles.

"Who is digging and who is building?" asked Uncle Tom.

Alex was keen to dig, whilst the others built the castle.

"What do you think happens to the sediment that rivers wash into the sea?" asked Alex.

"It sits at the bottom of the sea," said Joe.

"The sea is very heavy," said Alex. "So there must be a lot of pressure on top of the sediment."

"Does it turn to rock?" asked Sam.

"Yes, with enough pressure and time, it becomes what is called a sedimentary rock," said Uncle Tom, "like sandstone and shale."

"What about limestone," said Sam. Is that a sedimentary rock too?"

"Yes," said Uncle Tom. "Limestone is also formed under the weight of the oceans from the shells and bones that have settled to the ocean floor."

Alex had now dug so deep he could stand up to his waist in the hole.

Sam and Joe had used the sand to build a very grand castle with turrets and a surrounding wall. Joe carried several bowlfuls of water from the sea to try and fill the moat.

"I nearly forgot," said Uncle Tom. "I have your 4th challenge here."

He handed it to Joe, who read it out.

> Dear Alex, Sam and Joe
>
> After your successful completion of the third challenge, your fourth challenge is to find something to move the ocean
>
> Yours,
> Uncle Tom

"It must be something very strong," said Joe," it was hard enough just carrying those bowls of water back from the sea. And it was further to walk as the tide is going out."

"Oh, yes, the tides move the oceans," said Alex, "and the tides are caused by the Moon."

"But how?" asked Joe.

"The gravity of the Moon pulls on the Earth and the water, so the water bulges towards the Moon," said Alex.

"And the Earth is rotating," added Sam, "so that bulge of water moves around the Earth over the course of the day. And we see the sea rise and fall.

The time of tides will change each day because the Moon is also moving around the Earth."

"But there are two tides each day," said Joe, "where does the other tide come from?"

"I'm not sure," said Alex. "The tides are evenly spaced, so the second tide could be due to the effect of the Moon on the opposite side of the planet. We have just considered the water closest to the Moon."

"The gravity would pull on all of the Earth, but most strongly on the Earth closest to the Moon, so the ocean furthest away bulges outwards as the rest of the planet is being pulled more strongly towards the Moon," said Sam.

"Do you think there are tides on the Moon?" asked Joe.

"There is no water," said Alex.

"Yes, but the Earth will still pull on the Moon and would make the surface bulge," said Sam.

"Does the Moon rotate?" asked Joe.

"Yes," said Sam, "but the same face of the Moon always faces us. We always see the same face of the man in the Moon, the same craters. It just gets darker and lighter in parts over the month. So, the Moon is rotating at exactly the same speed that it goes around the Earth."

"Why is it exactly the same speed?" asked Joe, "or is it just by chance?"

"It is too exact to be a coincidence," said Sam.

"Maybe it's because of the tides," said Alex. "The tides on the Moon would be much stronger than on Earth because the Earth is much bigger, so its gravitational effect is stronger. If the Moon did rotate faster, the Earth would cause a tide of rock to move around the surface of the Moon as it rotated, so the Moon rotations would slow until the same face of the Moon was always facing and bulging towards the Earth."

"Wow, so gravity moves whole oceans and makes tides in rocks. That is very strong," said Joe.

"It is immensely strong," said Uncle Tom. "Well done, and here is another puzzle piece."

Alex took it and put in his pocket.

"Talking about time and tides," said Uncle Tom. "Your mum will be expecting us home soon. Let's get going."

# Chapter Nine

"You're late," said Alex's mum, "when they came through the door. I was worried where you had all got to. I've got your dinner ready for you."

They all sat down, hungry after the journey home.

"Sam and Joe, your mum called to say she will be home tomorrow."

"Oh," they both said in disappointment.

"You can come again anytime you know that," said Alex's mum, "but I think your mum will be looking forward to seeing you."

"But we haven't found the final piece of the challenges yet," said Joe. "Do you think we'll still have time?"

"I'm sure you will," answered Uncle Tom.

Alex's Mum tucked them into bed and Uncle Tom came up to tell them another bedtime story.

This story is about a person you may have heard of before, Albert Einstein.

"I've heard of him," said Alex, "he was a science genius."

"Yes, he was," said Uncle Tom, "he made some amazing leaps forward in our understanding of the universe. And he even looked like a crazy scientist with mad hair.

He was born in the late 1800s in Germany. As a child he was very interested in mathematics and how it could be used to explain the world. He grew up at a time of some great scientific discoveries, which he followed closely.

One discovery from the study of Electricity and Magnetism by Maxwell, is that light travels at a constant speed in a vacuum, such as space. (It does change speed if it is travelling through a substance such as air, water or glass).

It doesn't matter how fast you are travelling or in which direction when you measure the speed of light, it will always be the same speed. Even if you are moving at half the speed of light towards the measurement or away.

This is different from our everyday experience. If you measure the speed of a car when you are standing still, it would be moving away from you much faster than if you are sat in a car travelling along behind.

Einstein thought about this a lot and imagined what would happen if he was travelling along at the speed of light next to a light beam. He realised if the speed of light is always constant, then measurements of time and space must change when you are travelling at high speed."

"Really? Time and space can change?" said Alex.

"Yes, it is not something we notice as we never travel anywhere near the speed of light, but it is possible to measure by putting very accurate clocks on aeroplanes and when they land, the clocks onboard have measured less time than a clock that stayed on the ground."

"There is a famous twin paradox. If one twin stays on Earth and the other twin travels into space for several years then comes back to Earth, the stay at home twin would have aged more than the space traveller twin. Time runs slower at higher speeds."

"That's amazing, so if we travelled really fast we would stay young for longer?" asked Sam.

"Yes, but only relative to someone who wasn't travelling as fast. You wouldn't notice the time ticking by slower.

Einstein made this discovery just before World War I and carried on his research during the war. He considered what happens to light, time and space near the gravity of a large mass like a star."

"And what happens then?" asked Alex.

"He found that if you consider time as just another aspect of space, then gravity can be considered just a curvature of this "spacetime".

Mass bends spacetime. So instead of gravity being the pull of the Earth on another mass (such as you and I), we can consider the mass of the Earth bending the spacetime around it. Gravity is the effect we measure as a result of that curvature.

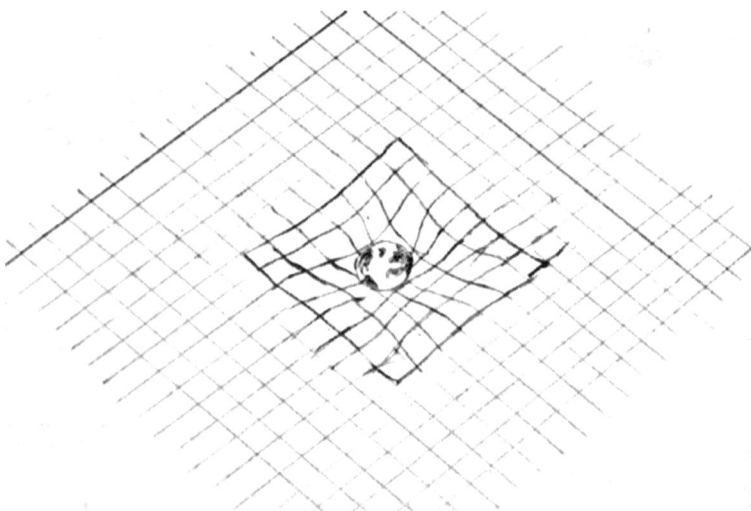

Just like if you sit in the middle of your bed, you make a depression in the mattress underneath you and if I rolled a ball onto the bed it would tend to roll towards you. The Earth (and any other mass) causes a depression in spacetime and another object that comes near moves towards the Earth. We can call this the gravitational pull of the Earth, but really it is just the natural movement of that object in the shape of the spacetime."

"So, what does that mean?" said Joe.

"It means even particles without mass such as light are affected by gravity. It also means that if gravity is very intense even light cannot escape from it, which we call black holes."

"What happened next?" asked Alex.

"It was the First World War and many English people did not want to discuss the work of a German scientist. However, there was one man, Arthur Eddington, who was the Astronomer Royal. He made it his personal mission to promote the work of Einstein. He believed scientific understanding was more important than nationality and wars."

"What did he do?" Sam asked.

"One of the predictions of Einstein is that light will be bent by the curvature of spacetime near a large mass, such as our Sun. But it is difficult to measure light from stars when it is near the Sun as the Sun's light is too bright to see them. But their light can be measured during a total eclipse when the Moon travels between the Earth and Sun and blocks out most of the sunlight so the stars can be seen."

"Eddington and his team travelled to Africa and Brazil to take photos of the stars during the eclipse to compare with how they appear when we can view them directly and they are not near the Sun.

The effect is tiny, but it was measurable, and Einstein became world famous.

It was much later that his predictions regarding black holes were also proved to be true."

"Does gravity affect time as well?" asked Sam.

"Yes, it does. If we measure time when gravity is stronger, it appears to run slower than measurements where gravity is weaker. So your head is actually older than your feet."

"Do astronauts age more quickly when they are in space?" asked Alex.

"The lower gravity would suggest they should age quicker, but they are also moving faster which would slow their ageing. Overall, they actually age slightly slower than if they stayed on Earth. The effects are tiny. But they are important, not for humans in orbit, but for the satellites around Earth such as those used for our navigation and communications which rely on exact timing.

Now I think it is time for sleep. Sweet dreams."

Alex lay awake imagining how spacetime curved and time could run faster and slower. It made his head spin and finally, he closed his eyes and dreamed of riding on a light beam through the stars.

# Chapter Ten

Sam and Joe woke up early. They were going home today and wanted to make the most of the time before their mum came to collect them.

"Wake up Alex," said Joe, patting him on the shoulders.

Alex slowly opened his eyes.

"Time to get up," Joe added. "We need to finish the final challenge."

They ran downstairs to check if there was another envelope from Uncle Tom.

"You're up early," said Alex's mum. "Are you looking for your next challenge, Uncle Tom left you this."

She handed them an envelope.

"Now sit down and have some breakfast whilst you read it."

Sam opened the envelope and read out the letter.

> Dear Alex, Sam and Joe
>
> Congratulations on completing four challenges. Your final challenge is to grow older quicker.
>
> Yours,
> Uncle Tom

"I think from what Uncle Tom told us last night, we need to move more slowly, or move to where gravity is weaker," said Alex.

"I don't think we can move more slowly since we are sitting still, "said Joe.

"So, we need to go to where gravity is weaker then?" said Sam. "Hang on, there is a little key. Do you have the puzzle pieces we have collected so far?"

"Yes," said Alex, and tipped them onto the table.

"It looks like a map," said Joe.

They arranged and rearranged the pieces and finally, they could see a sketch with a cross.

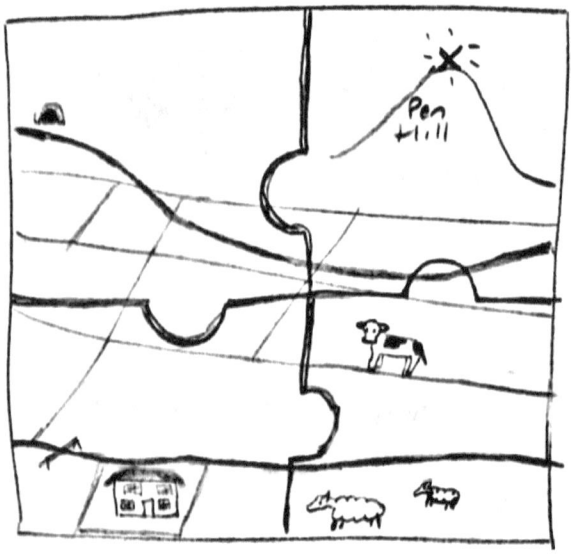

"It is a map," said Joe, "look there's the house."

"And there is a cross marked on the top of Pen Hill," added Sam. "Let's go."

"We're just going for a walk up Pen Hill," said Alex, "we'll be back for lunch."

"Don't be late, Sue is coming this afternoon."

"We won't," replied Alex.

They set out, running up the track towards Pen Hill.

They slowed down as the path became steeper.

"What do you think we will find?" asked Joe. "Do you think there will be treasure?"

"I'm not sure, but we'll find out soon," replied Alex

It was cloudy today and the wind felt cold. "Do you remember Uncle Tom telling us about climbing Everest? I'd never really appreciated how important our atmosphere is and what it would be like if gravity was just a little bit weaker," said Sam.

"It would be a very different world," said Alex, "there would be no tides, the boats wouldn't sail on the water."

"I guess we wouldn't have oceans, as the rivers wouldn't flow down to the sea. And without air pressure the water would boil away," added Sam.

"If there was no gravity, we wouldn't have a Sun or Moon or atmosphere at all. In fact, the Earth and planets would not have formed. It's gravity that gives us the structure of the universe, the stars, planets and galaxies," said Alex.

They were almost at the top and they could see the glint of something on the summit.

"That must be what we are looking for," said Alex, "come on."

They ran up the final stretch and saw a little golden chest at the very top.

"Have you got the key?" asked Sam.

Alex took the key from his pocket and gently tried it in the lock, it turned easily.

Opening the lid, there were 3 golden medals on golden ribbons.

"Wow, they must be for us," said Joe and eagerly took his.

There was a rocket on one side with the planets in the background.

On the other side was a big question mark with words around the outside.

"PhysWizz Gravity Award," read out Sam.

"Well done," said Uncle Tom as he came up from the other path. "You guys did a fantastic job. You can be really proud of yourselves and you more than deserve your medals.

Just remember to keep questioning and you will be amazing scientists."

They put their medals around their necks, and all walked back home together.

"Thanks, Uncle Tom," said Alex. "That was great fun. Will there be more challenges soon?"

"I'm sure there will," said Uncle Tom.

Lunch was on the table when they got back.

"I've got an extra surprise for you," Mum said and lifted a question mark cake out of the oven. "I'm very impressed with how you questioned and worked out the answers to the challenges."

Sue arrived just after lunch to pick up Sam and Joe.

"Did you have a good time?" she asked.

"Yes, it was great, can we come again?" replied Sam

"Well, that depends on Helen," said Sue.

"They are welcome any time. You know that. Alex loves having them here," said Helen.

"Hurray!" said Joe, "I can't wait for the next challenge."

*"Because there is a law such as gravity, the universe can and will create itself from nothing."*

*Stephen Hawking*

# About the Authors

Ros Greener studied Natural Sciences at Cambridge University, specialising in Theoretical Physics and receiving a Class I degree. She is interested in changing the perception of science as something hard to understand into common knowledge, and especially the teaching of science to young children. She believes if scientific understanding can be embedded at a young age, children will grow being able to reason and make sense of the world they experience and will learn through their own questioning.


Alexander Greener is fascinated by the scientific explanation of everyday observations and how scientific knowledge can be used to make new machines and inventions. He has a passion for writing and reading, as books allow the reader more freedom to use their imagination than other formats.

# Visit www.PhysWizz.co.uk

For details of other titles and more content.

www.ingramcontent.com/pod-product-compliance
Lightning Source LLC
Chambersburg PA
CBHW021848170526
45157CB00007B/2991